HOW CAN I EXPERIMENT WITH ... ?

AN INCLINED PLANE

David and Patricia Armentrout

Rourke
Publishing LLC
Vero Beach, Florida 32964

www.rourkepublishing.com

PHOTO CREDITS:
©Armentrout pgs 4, 8, 14; ©David French Photography pgs 19, 20, 23, 27; ©Image 100 Ltd. Cover, pg 16; ©Digital Vision Ltd. pg 25; ©Jonathan Kirn/Stock Connection/PictureQuest pg 13; ©PhotoDisc pg 7; ©Painet, Inc. pg 29.

Cover: *A flat sloped surface can be a fun ride!*

Editor: Frank Sloan

Cover design: Nicola Stratford

Series Consulting Editor: Henry Rasof, a former editor with Franklin Watts, has edited many science books for children and young adults.

Library of Congress Cataloging-in-Publication Data

Armentrout, David, 1962-
   How can I experiment with simple machines? An inclined plane / David and Patricia Armentrout.
      p. cm.
Includes bibliographical references and index.
   ISBN 1-58952-333-4
   1. Inclined planes—Juvenile literature. I. Title: Inclined plane.
II. Armentrout, Patricia, 1960- III. Title.
   TJ147 .A759 2002
   621.8—dc21
                                                            2002007951

Printed in the USA

w/w

Table of Contents

Inclined Plane (IN klynd PLAYN) — a flat sloped or slanted surface; a simple machine used to make work easier

An inclined plane makes the job of unloading a truck easier.

Machines

People use machines because they make work easier. An inclined plane is a fancy term for a simple machine. A simple machine has very few parts. Some simple machines have no moving parts, like the inclined plane. An inclined plane is nothing more than a flat sloping surface that makes work easier.

The wheel, the pulley, the wedge, the screw, and the lever are simple machines, too. If we put simple machines together, we can build complex machines.

This rail car is called an incline. It shuttles people up and down a steep hillside.

How We Use Simple Machines

Wheels reduce friction. Pulleys help lift heavy loads with less effort. Screws can hold things together. A lever makes moving heavy objects possible.

Think about the car—a common, but **complex**, machine used everyday. Look for simple machines at work. First, you may spot the wheels. There are four on the ground and a steering wheel inside. However, there are gears in the engine. Did you know that gears are wheels, too?

Levers are used to open doors. Screws hold parts of the car together. Can you find other simple machines at work?

A car is made up of many simple machines. Can you spot any here?

Understanding the Inclined Plane

The inclined plane is a slope, but how does that make it a machine? How does a slope make work easier?

Imagine you are a delivery truck driver. Your job is to load and unload refrigerators. The problem is that the back of your truck is 4 feet (1.2 meters) off the ground. Can you lift a refrigerator 4 feet (1.2 m)? Probably not. An easier way would be to use a ramp. A ramp is an inclined plane. A ramp gives the driver a mechanical advantage. A **mechanical advantage** allows the driver to use less effort to move heavy objects in and out of the truck.

Without a ramp, these men need to do a lot of heavy lifting.

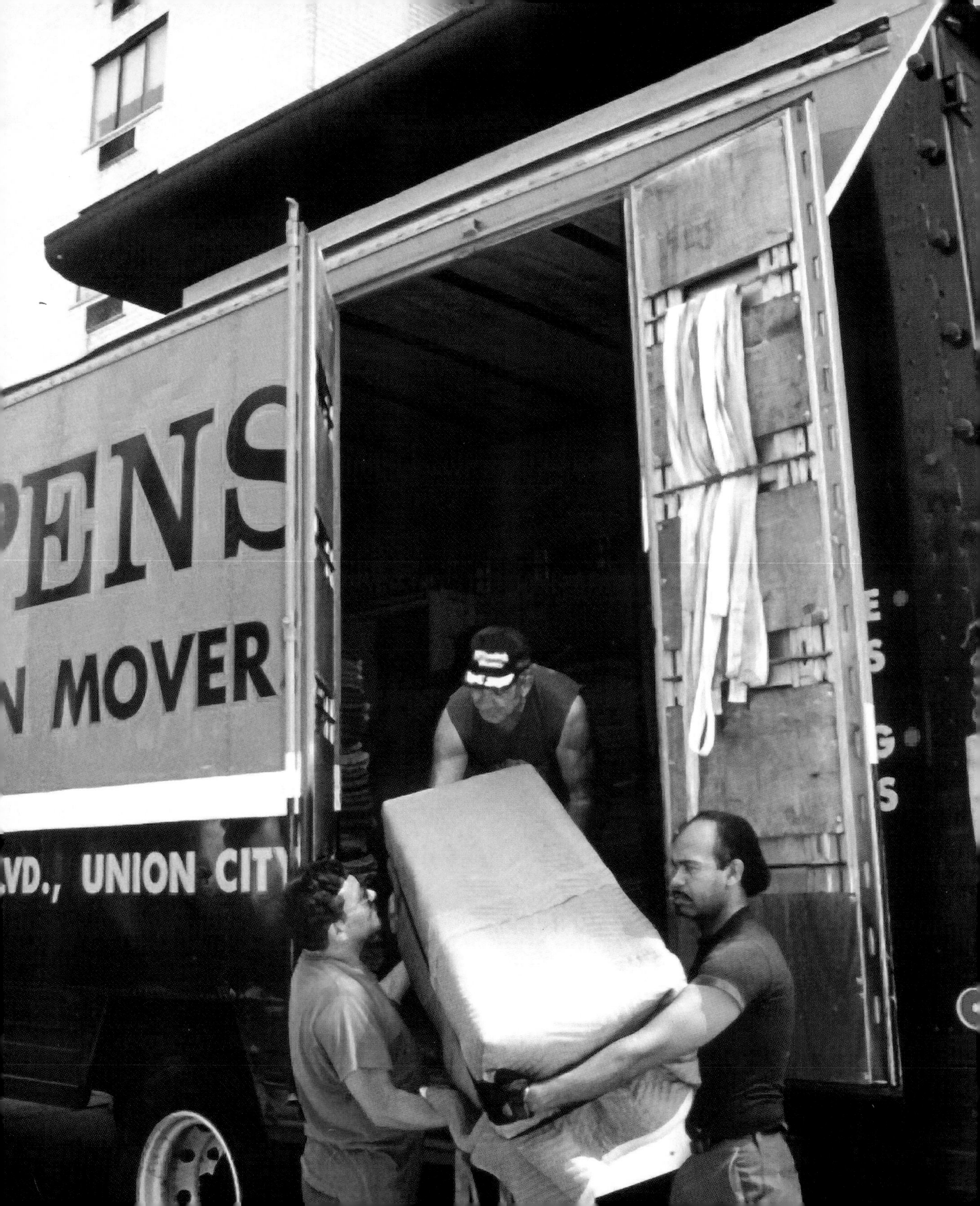
PENS
N MOVER
LVD., UNION CITY

Work

Work can be defined using a formula:

$$FORCE \times DISTANCE = WORK$$

To understand this formula, use the example of the delivery driver. If the driver uses a short ramp, the **distance** he travels is also short. A short ramp is steep. The driver would need to use a lot of **force** to push a refrigerator up a steep ramp.

If the driver uses a long ramp, he must push his load further. A long ramp is not as steep. The driver would not need as much force to push his load up the ramp. The driver uses less force, but over a greater distance. Whichever ramp he uses, the amount of work is the same.

Furniture movers prefer a long, gradual ramp to a short, steep ramp.

All Shapes and Sizes

Inclined planes come in all sizes. A step stool is a small inclined plane that helps us reach high places. A road that works its way up the side of a mountain is a big inclined plane.

It is easy to see why we need different sized inclined planes. Size is important, but what about the angle of the slope? The angle is important, too. The angle makes the slope steep or gradual.

Would you rather climb a short steep slope or a long gradual slope? Most of us would rather take the long gradual slope because climbing the short steep slope is harder.

A gradual slope on a playground slide is easy to climb.

More about Inclined Planes

Inclined planes work because of **gravity**. Gravity is the force that holds objects on Earth. Without gravity, an inclined plane would not be needed. Inclined planes help us by supporting the weight of the object we are moving while we push or pull it up the slope. Since we don't have to hold the entire weight of the object, our job becomes easier.

Inclined planes do not do all of the work for us. They just allow us to spread our work over a longer distance.

Gravity is the force that pulls all objects toward the center of Earth.

Lifting a Load

You will need:
- **tape**
- **4 feet (1.2 m) of string**
- **small toy car**
- **adult**
- **empty film container**
- **bench**
- **handful of pennies**

Tie or tape one end of the string to the car. Have the adult punch a small hole into the side of the film container close to the top.

Put the free end of the string through the hole in the container and tie a knot. Place the car on the floor next to the bench. Run the string over the top of the bench so that the weight container hangs just below the opposite edge. Adjust the string length if necessary.

Drop pennies into the container one by one until the car reaches the top of the bench. Record the number of pennies you used to lift the car.

How many pennies will it take to lift the car off the floor?

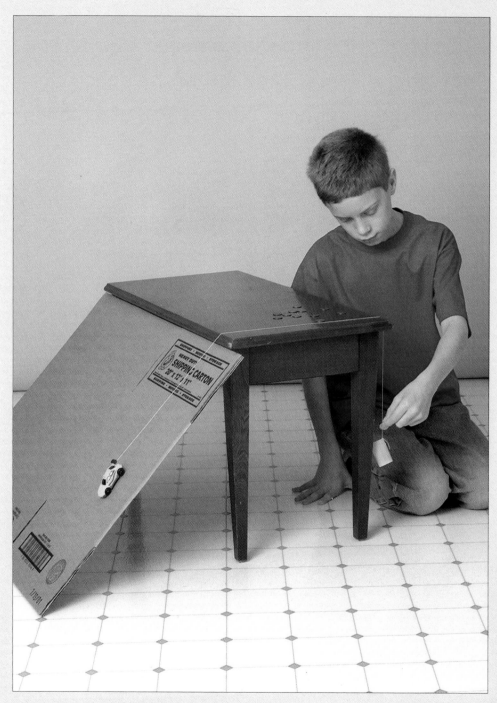

Add an inclined plane to help lift the car off the floor.

Lifting a Load Using a Steep Inclined Plane

This time add an inclined plane and do the experiment again. You will need a 2 x 3 foot (0.6 x 0.9 m) piece of stiff cardboard for your inclined plane.

Lean the cardboard so the long sides rest at the top of the bench and on the floor. Place the car on the bottom of the plane with the string running up and over the top of the bench. Your weight container should hang just below the opposite edge. Adjust the string length if necessary.

Drop pennies into the container until it drops to the ground and the car is pulled up the slope. How many pennies did you use to lift the car on your inclined plane?

Lifting a Load Using a Gradual Inclined Plane

What would happen if you had a longer inclined plane? Turn the cardboard so the short sides rest at the top of the bench and at the floor. Notice how the angle has changed. Do you think you will need more or less weight to pull the car up the slope? Try the experiment again.

The same amount of work was done in all three experiments—the car was raised to the same level. The number of pennies represented the amount of force used to raise the car. Did you need more or less force as the slope got longer?

It takes less force to lift a car when you use an inclined plane.

23

Friction

Two objects rubbing against each other cause **friction**. Friction is a force that slows movement. If you slide a book across the floor, eventually the book will stop. What stopped the book's movement? Friction.

Long ago, people discovered that wheels reduce friction. Wheels used with the help of an inclined plane make it easier to move loads. In the earlier experiments, you found that an inclined plane allowed you to use less force to raise the toy car. Would the results have been different if the car did not have wheels?

Wheels allow bike riders to glide easily on pavement.

Experiment with Friction

Without wheels, the car would have created more friction. More force (pennies) would have been needed to raise the car up the slope. Let's see how much difference friction makes.

You will need:

- **two toy cars of about the same size**
- **stiff piece of cardboard**

Place the cars next to each other at one end of the cardboard. Turn one car upside down on its roof. Slowly raise the end of the cardboard where the cars are.

Which car begins to roll or slide down the inclined plane first?

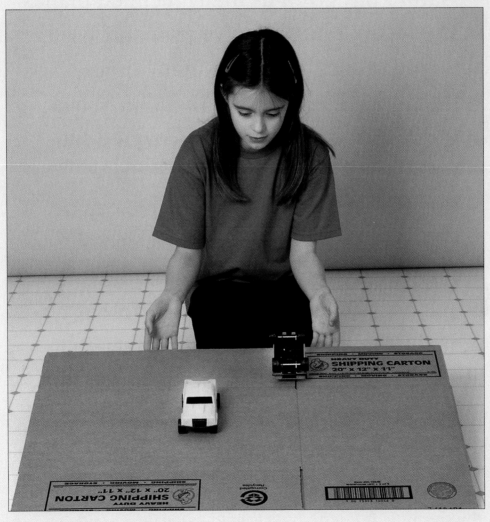

Wheels reduce friction.

Types of Inclined Planes

Nature has given us many inclined planes in the form of hills. You've probably used this type of inclined plane many times without giving it a second thought. When you ride a sled down a snow-covered hill, you are using a natural inclined plane.

Inclined planes made by people can be found almost anywhere. Ramps, slides, escalators, and staircases are great examples of inclined planes. Notice the different types of inclined planes people use as you go about your day. You may be surprised at how many you can find.

A water slide is a refreshing way to enjoy an inclined plane.

Glossary

complex (KAHM pleks) — made up of many parts

distance (DIS tens) — the amount of space between two points

force (FORS) — the push or pull needed to move an object

friction (FRIK shun) — a force that slows two objects when they are rubbed together

gravity (GRAV eh tee) — the invisible force that pulls objects to Earth

mechanical advantage (mi KAN eh kul ad VAN tij) — what you gain when a simple machine allows you to use less effort

Further Reading

Macaulay, David. *The New Way Things Work.*
  Houghton Mifflin Company, 1998
VanCleave, Janice. *Machines.* John Wiley and
  Sons, Inc., 1993

Websites to Visit

http://www.kidskonnect.com/SimpleMachines/
  SimpleMachinesHome.html
http:www.mos.org/sin/Leonardo/
  InventorsToolbox.html
http://www.brainpop.com/tech/simplemachines

Index

About the Authors

David and Patricia Armentrout have written many nonfiction books for young readers. They specialize in science and social studies topics. They have had several books published for primary school reading. The Armentrouts live in Cincinnati, Ohio, with their two children.